SEND ME

STEVE KELLY

with Joe Riddle

Printed in the United States of America.

ISBN 978-1-940717-16-6

DEDICATION

This book is for the men, women, and their families who have answered the call to defend this nation. We owe you a debt we can never repay.

Table of Contents

Here am I. Send Me.

Isaiah 6:8

INTRODUCTION: Who Will Go For Us?

Then I heard the voice of the Lord saying, "Whom Shall I Send? And who will go for us?" And I said, "Here am I. Send Me."

- Isaiah 6:8 (NIV)

When a service member takes the Oath, they swear to "support and defend the Constitution of the United States against all enemies, foreign and domestic..."[i] You promised to faithfully oppose any threat endangering our nation and its way of life.

The United States Armed Forces are proudly a volunteer military. This means nobody forced you to enlist. You stepped up because you saw the need. You heard the call, and you took action by joining the greatest fighting force in modern history.

Through this single act, you proceeded to write a

blank check. However, this check is different from any other. This one was endorsed with your very life, made payable to the people of the United States of America.

In this sense, both the prophet Isaiah's commission, seen in the verse at the beginning of this section, and your own are quite similar. Both are entered into willingly as an act of free will.

This book was written to help you realize that the same moral code guiding you in your national service was first and foremost the ethos of heaven, and it still is.

Following Christ is not for the faint of heart. It requires a warrior's heart. Jesus asks for all or none of your devotion.[ii] While this may seem very black and white, there is plenty of room for God's grace. C.S. Lewis described this life quite interestingly when he wrote the following:

> "Enemy occupied territory—that is what this world is. Christianity is the story of how the rightful king has landed… and is

calling us to take part in a great campaign..."[iii]

God is looking for men and women who will stay their post. Heaven's conflict is not waged along geo-political lines. This means the prize is not access to water, a new system of governance, or earthly power. This war is unlike any other. It is a no holds barred fight for the heart and soul of all mankind. Eternity, and where people will spend it, is at stake!

Being in the military often means you cannot say with great certainty where you will be in a month, much less two or three years from now.

You should know that no matter where your orders take you, the call to aid in the cause of the Kingdom follows you. Whether you are a sailor aboard a sub deep in the heart of the ocean, an airman in the upper edges of the atmosphere, a marine in a conflict zone, or a soldier in faraway lands: Heaven asks the question, *"Will you go for us?"*

Over the years, some service members in our church have struggled with reconciling their faith and military service. It's as if they felt God can't use a person who has seen what they have seen, and done what their job has required them to do. Nothing could be further from the truth! Your call as a service member in our armed forces is unquestionably part of God's plan.

Paul writes in Romans 13, "Let everyone be subject to the governing authorities, for there is no authority except that which God has established."[iv] What is Paul saying? Governments and militaries are divinely raised up by God for His purposes.

In Acts, we read about Cornelius, a Roman centurion in charge of one hundred soldiers. He's described as a "devout man, and one who feared God."[v] At no point in his conversion experience does Peter question the morality of his career as a soldier.

In Exodus, Moses writes, "The Lord is a warrior. The Lord is His name."[vi] Make no mistake about it: our God feels at home amongst warriors. He saw to it that He would be identified as one in scripture.

Both the military and the life of a Christian are about choosing service over self. They both call you to run to the front lines, while most are running away. Isaiah's willingness to answer Heaven's call should challenge us to do the same.

He volunteered without the security brief. He stepped up without a topographical layout of the target area. He agreed to a mission without knowing all the details. He had the one element that mattered most. He knew there was a need.

Heaven is calling. Will you answer? Answering the call of Christ is not always safe or comfortable. That said, our God is a warrior. He will not be contained. He willingly invites each of us to be a part of what He is doing if we so choose. Will you go?

Chapter One: ALWAYS FORWARD

Make no mistake, there is a fight for your focus. To say focus matters would be an understatement. The reality is that your focus forms your future. With this truth in mind, it should come as no surprise that the writer of Hebrews commands the reader to "fix your eyes on Jesus, the pioneer and perfecter of our faith."[vii] If we are going to step into everything Christ has for us to accomplish on this earth, we must fight to keep the proper focus.

I pastor Wave Church in Virginia Beach, Virginia. I'm proud to tell people I'm from here! Not everyone can say that about where they live. One of the main reasons I'm so proud is because our city is home to some of the most revered warriors on the planet, the U.S. Navy SEALs.

I quickly realized nothing about these men is average. This is even true about their focus.

After a few lunch meetings with operators, I realized they pay attention to every detail. No matter the topic of discussion, these guys always seem to have a read on what's happening around them. It's what they call "keeping your head on a swivel". To say that they are aware of their surroundings is an understatement. They are always ready for anything to walk through the door at any time.

It wasn't long before I realized there was something to this type of behavior. Warriors realize you may not always get to choose when the battle begins, but you do choose the level of your preparedness.

In this life, it is not a question of if the battle will come. Your day of battle is en route. The question is will you be focused and ready when your time comes? It is that reality which brings me to an essential question. Are you living focused on the things that truly matter?

The title of this chapter, *Always Forward,* is derived from the motto of the United States

Marine Corps Forces Special Operations Command (MARSOC). This motto is indicative of what separates the proverbial "sheep dog" from the "sheep".[viii]

The average person takes the time to lick his wounds. The average individual makes the most of every opportunity to rehash the wrongs life has dealt them. The warrior is different. He or she does not have the luxury of thinking this way.

Your opposition does not wait until you are ready. Instead, it actively seeks out those areas of unpreparedness in your life. The enemy searches for the gaps in your defenses and intends to take full advantage of them.

When the attack comes, the average person is knocked down in the ambushes of life and immediately goes into flight mode. Not the warrior. The warrior chooses to fight.

The day will come when we get knocked down: in our careers, in our walk with Christ, in our marriages, in our parenting, or virtually any other arena of life.

Perhaps it feels like someone in your unit has made a point to soil your reputation. It's not over. Fix your focus. Always forward. That is the way of the warrior. Sulking never changes anything; you have ground to take in life!

A riveting scene in the movie *Saving Private Ryan* happens when Tom Hank's character, Captain John Miller, is storming Omaha Beach on D-Day. In the midst of all the chaos, one of his soldiers looks to him. Surrounded by death and the air thick with lead, the Ranger asks Capt. Miller, "What's the rallying point?"

Capt. Miller immediately screams, "ANYWHERE BUT HERE!"[ix] I've read a number of personal accounts of what that day on the shores of France was like. Time and again, I've heard that as bad as it was, not taking ground was never an option in the minds of those warriors that day.

Your season may be difficult. Your pursuit of Christ may be costing you something even now as you read the text on this page. That doesn't excuse you from the need to take ground. Warriors are either up or getting back up. Take God at His word!

Sitting still on the beaches of Normandy was a sure death sentence. In the same way, sitting still in the midst of your life battles is not a viable option either. Always forward!

Where in life have you stopped moving forward? Where in life have you become comfortable? God's warriors must develop an "Always Forward" mentality built solely upon trust in God.

If you trust your skill, your strength, your cunning, or your own intellect then you will come up short. When we trust God, His unparalleled sufficiency becomes our strength and gives us total dominance in the battle spaces of life.

If we are truly "Always Forward" followers of Christ, we need to acknowledge there will be times we want to quit. We must daily decide not to tap out. Staying in the fight doesn't mean it's easy. Life has more in common with a battle than it does a pleasure cruise. If you want to live to tell the story, it requires a commitment that says, "I will succeed or you will drag me away in a body bag." As followers of Christ, we must have the mentality which says, "I'm playing for keeps. Heaven and Hell are real. The stakes are eternal. I will not back down. Always Forward!"

I'd like to tell you developing this mentality means you will always feel up for the fight. You won't. You won't always feel ready, or like you belong in all the situations you find yourself in.

In my thirty plus years of ministry, there have been a number of occasions where I honestly felt I was in over my head… and that is ok. As a warrior, you can feel any emotion you want. However, you cannot allow it to have the final say and take control of you.

When fear or doubt knock at your door, you need to remind yourself of the timeless wisdom of Isaiah 40. In the midst of some tough days for the people of Israel, God sent these words through his prophet Isaiah:

> "He gives strength to the weary and increases the power of the weak. Even youths grow tired and weary, and young men stumble and fall; but those who hope in the LORD will renew their strength. They will soar on wings like eagles; they will run and not grow weary, they will walk and not be faint."[x]

Such great encouragement can be found in those words! They reveal to us that God's aim is not to sap your strength. Instead, He desires to renew it.

Perhaps this book has found its way into your hands in the midst of a deployment and things are difficult. It could be that the battle for your personal purity is in a state of total war.

Maybe your squadron, platoon, or company has been dealt a few bad hands. Maybe you have found hesitancy creeping into your heart. I don't know your current state as you read this, but I do know my God. His heart is to renew your hope, and move you forward. His desire is that you would "walk and not grow faint!" Always Forward!

As long as you have breath in your lungs, you can rest assured that there is more in store for you to accomplish.

I always put it like this to our church: "The best is yet to come!" How can I be so certain? Because Jesus is for us and not against us. Do you truly believe this is true for you right now? Do you know that God isn't mad at you? Do you believe Jesus desires to walk in personal relationship with you? Do you understand that the Holy Spirit is waiting for you to give Him room to walk and talk with you in your daily life?

Don't doubt the fact that God can and will use somebody exactly like you for His purposes. His heart for you is "Always Forward!"

When God is truly your strength, what excuse do you have to not move forward in life? David, a warrior king to the people of Israel, wrote, "The Lord is my strength and shield; my heart trusts in Him and He helps me."[xi]

You don't need to hide from God. No matter the situation, in the midst of whatever it is that you face, you can turn to Him. I've learned that regardless of the size of the storm, it's never a good enough reason to stop taking ground.

Is that because I doubt the size of your obstacle? Not at all. I say that with such confidence because I've come to know the size of the God I serve. In Jesus, you have an unlimited source of strength to help you develop an "Always Forward" mentality.

If I could impart to you a key skill for developing an "Always Forward" mentality, it would be this: know what the Bible has to say about the strength of our God. Before every Marine goes to war, he or she must learn *Marine Corps Doctrinal Publication One* inside out.

In the same way, I believe every Christian must

know their Bible. For many of us, the reason we lack an "Always Forward" mentality in life is because we have a comprehensive understanding of our opposition but we do not possess the same depth of understanding about our God and His Word.

Invest time in reading your Bible. It pays dividends in every area of your life. Habakkuk 3:19 says, "The Sovereign LORD is my strength; He makes my feet like the feet of a deer, He enables me to tread on the heights."[xii] The devil would love for you to think that being a Christian in an environment like the military would only serve to weaken your reputation and ability to perform your duties in the service of our nation.

Nothing could be further from the truth! As we embed ourselves in a local church and study the word of God, it will position each of us to "tread on the heights."

Army Regulations 670-1 states that the U.S. flag patch is to be worn so that

> "the star field faces forward, or to the flag's own right. When worn in this manner, the flag is facing to the observer's right, and gives the effect of the flag flying in the breeze as the wearer moves forward."[xiii]

The flag is always facing forward as if charging in to battle. Always Forward, Never Retreat. You were created to take ground in life. The next time you put on your uniform and see that flag facing forward, allow it to remind you of the man or woman you are called to be in Christ. My prayer for you is that your life could be described using the motto, Always Forward!

Chapter Two:
THAT OTHERS MAY LIVE

Warriors are a different class. The days others fear are the days warriors live for. They passionately pursue what most aggressively avoid. What is at the core of this behavior? Is it an inability to properly assess and know the dangers that are so clearly evident to everyone else around them? Is it youthful recklessness? Hardly.

The theologian G.K. Chesterton insightfully wrote "the true solider fights not because he hates what is in front of him, but because he loves what is behind him."[xiv] Therein lies one of the core values of the warrior: love for another. This willingness to put others ahead of oneself is a prerequisite of the warrior ethos. A valuable question we must all ask ourselves is "what has my heart?"

The United States Air Force Pararescue (commonly called PJs) has a unique mission profile:

"When an Airman needs saving, it is these highly trained experts' duty to do everything possible to bring them home."[xv] There is no fine print to this mission set. These brave men have been called upon many a time over the years to knowingly put themselves in harm's way in order to give another a fighting chance at survival. No matter the cost.

The motto of this highly-decorated unit in the American Special Operations Community clearly communicates their self-defined reason for existence: *"These things we do that others may live."* In a world obsessed with self-promotion, they have chosen the road less traveled.

They do not serve in search of honor. They do not serve in order to preserve their own lives. They do what they do so others can live another day. Their example boldly reveals one of the essential characteristics of the people God uses: warriors possess a deep rooted commitment to others.

Jesus was teaching His disciples when he said, "Greater love has no one than this: to lay down one's life for one's friends."[xvi] These words underscore a crucial difference between the Kingdom of God and the Kingdom of Self. It's not about you! Simply nodding our heads in agreement with this idea is easy. The reality is that living it out may very well cost you everything.

Here in the United States, we are encouraged by society to make life all about ourselves. While I don't think God has any issue with us living blessed, we cannot allow our blessing, sense of calling and life focus to be all about ourselves.

Ours is the world full of self-serving electronic devices, coffees created to our exact preference, and social media that leaves us not so social. In this type of world, a dangerous shift can quickly and easily happen in our hearts before we even recognize it.

I believe that when we lose our focus on others, we lose the drive that enables us to do what warriors are called to do.

Jesus was able to serve and love like nobody else because His focus was never on Himself. He lived life driven by the dream of connecting a broken humanity with its Creator.

Never forget, God desires to use you in the season you are in right now! You don't have to wait until your next promotion. God's plans can happen with or without a new set of orders. His call on your life is not on hold until you are out of the service.

In the midst of the season you are in, Jesus has great plans to reach the world around you… through you. But to accomplish all He has for us to do, we must cultivate a selfless heart that says, "I will lay down my life so others can find their own in Christ."

After a long mission, the soldier must immediately clean his gear. This insures his kit will be ready should he need to respond to a threat at a moment's notice. It's unthinkable to allow one's weapon to not be battle ready in a time of war. War does not stop so you can nap.

The most valuable tool in the Christ follower's arsenal is the heart. We must constantly cultivate a healthy heart so that we can be ready in and out of season to preach the Gospel to those in the world around us.[xvii]

In light of this truth, ask yourself: how ready is your heart for the day of battle? Have you allowed everything to be about you? Have you allowed the politics of your career in the military to eclipse the value of your integrity and testimony? When is the last time you freely accepted less than you deserved so others could receive more than they deserved? We are most like Jesus when we put the needs of others ahead of our own.

Paul was instructing his younger follower, Timothy, when he said,

> "Join with me in suffering, like a good soldier of Christ Jesus. No one serving as a soldier gets entangled in civilian affairs but rather tries to please his commanding officer."[xviii]

As Christians, our commanding officer is Jesus. I've found one of the most incredible ways to transform your reading of scripture is to stop treating the words of Jesus as the weak *requests* of a passive individual. Instead, see them as the *commands* of an officer preparing to lead you into war.

Seeing His teachings in this manner will better equip and position you as a warrior fighting for the cause of Jesus.

I encourage you to spend some time thinking about how genuine your commitment to living for others truly is. What motivates your leadership? Self-preservation or self-sacrifice? Until we gain a deep-seated commitment to living for others, we cannot become the brave-hearted and principled warriors God is calling us to be.

Just as a soldier's commitment to his brothers never reaches a limit, so too the Christian never arrives at a place where he or she can afford to allow life to become about self.

John the Baptist was preparing the hearts of his hearers for the coming of Jesus when he said, "He must become greater and I must become less."[xix] The enemy would love to convince you that living a selfless life diminishes your potential for success. He wants you to think that you would become weak, living a small and insignificant life. Nothing could be further from the truth! *You never lose by saying yes to Jesus.*

In letting go, we gain so much more. There is a call to a life of great impact awaiting you! However, it will never be found in the blind pursuit of opportunity. It's found through planting ourselves firmly on a foundation of principle.

My prayer is that you would answer the call to live a selfless life today!

Chapter Three:
WE OWN THE NIGHT

The purpose of this book is not to simply identify the traits of a warrior. Our aim is to challenge and equip you to live them. The world is filled with people who hold the ideals of the warrior high on a pedestal. It makes for good movies, great books, and cool guy clothing. However, when the storms of life hit, things will quickly fall apart for many.

The way of the warrior is more than chest beating and shouting. The truth is anybody can be brave when the odds are in his favor. The warrior proves him or herself through rising above the seemingly unbeatable odds and staying his or her post.

Heraclitus, a well-known Greek philosopher, famously said,

> "out of every one hundred men, ten shouldn't even be there, eighty are just targets, nine are the real fighters, and we

> are lucky to have them, for they make the battle. Ah, but the one, one is a warrior, and he will bring the others back."[xx]

The sobering reality is genuine warriors are sorely needed in the Kingdom of God but glaringly absent from the battle lines. Jesus speaks about this clear and present danger when He said "the harvest is plentiful, but the workers are few."[xxi]

Reflect on the words of Heraclitus and Jesus for a moment. Honestly ask yourself, which one are you? Are you the "warrior" Heraclitus was speaking of? Are you the "worker" Jesus had in mind—a person whom He can entrust His Kingdom to and also one whom He can build upon? Perhaps one of the most honest and accurate assessments of if you are truly battle ready is how you handle the valleys and the dark spaces of life.

The 160th Special Operations Aviation Regiment (SOAR) is a special helicopter aviation force in the United States Army.

Their missions include attack, assault, reconnaissance and more. One of the things that sets them apart is their incredible proficiency in night operations. At high speeds and low altitude… literally just feet above the surface of the earth. This unit thrives on the missions that give planners little or no notice before orders need to be executed.

Some of the United States military's finest aviators serve in this elite, secretive aviation regiment and have had key roles in a myriad of missions. Many of these exploits will likely never be known to the public due to their level of sensitivity. Their missions are state secrets. The motto of this particular community is "Night Stalkers Don't Quit!"[xxii]

These brave individuals are able to do the seemingly ridiculous in their field for no other reason than they don't know the meaning of impossible. This is not to say they don't comprehend it. They just won't agree to it being an acceptable outcome. If you can dream it up, they'll find a way to make it happen.

The only way they will ever use the word "quit" is to explain what they refuse to do. Quitting may be an option for some. This community has decided they simply will not do it.

As previously stated, one of their primary capabilities is flying at night. The advancement in technology enabling these aviators to fly in the darkest of nights with great precision is night vision.

Have you ever seen someone put night vision goggles on for the first time? At first, they stumble around and seem completely uncertain of their next step. On the other hand, when you see a person properly trained, it is amazing to see them walk through the darkest spaces as if it were broad daylight.

Imagine the 160th delivering troops on the doorsteps of America's enemies and you'll begin to understand why they confidently say a well-known Army motto: "We own the night!" For thousands of years, darkness posed an incredible challenge to strategists and warriors alike. How can you wage war without being able to see?

The Night Stalkers have proven their dominance many times in battle spaces around the world. Their ability to continue operating in the midst of the darkest night with the most demanding expectations imaginable has truly made them a fighting force without peer.

Anybody can stand for Jesus when it's popular. Anybody can be generous in seasons of financial surplus. It doesn't take great character to stand up for what's right when what's wrong isn't an attractive offer. But what about the night seasons? When trying times come your way, will you stand?

King David, a mighty warrior in his own right, wrote Psalm 91. It has been called "The Soldier's Psalm," and if you have not studied it in depth, I encourage you to do so. There is tremendous strength and wisdom for the warrior in those 16 verses. In this Psalm, David writes,

> "Whoever dwells in the shelter of the Most High will rest in the Shadow of the Almighty. I will say of the Lord, 'He is my refuge and my fortress, my God in whom I trust…' He will cover you with His feathers, and under His wings you will find refuge; His faithfulness will be your shield and rampart. You will not fear the terror of night, nor the arrow that flies by day, nor the pestilence that stalks in the darkness, nor the plague that destroys at midday."[xxiii]

King David teaches that a warrior right with God can access the confidence to tackle any season. As many reading this book would know, the theories of friction and fog of war are very real.[xxiv] When we hit a night season it's very easy to begin questioning everything.

Before we know it, we can find ourselves sapped of all courage and confidence. In just a moment, life can easily shift from the offensive to the defensive. What once felt effortless suddenly feels impossible to accomplish.

Have you ever been there? Perhaps you are there now. I know I've found myself in that position many times. How was David able to be so confident in his night seasons?

I believe the answer is hidden in an old training adage: train as you fight! David did not wait until the day of battle to know His God. Just as the Night Stalkers don't wait until the mission arrives to learn to fly in the night, we cannot wait until the tough times to learn how to rely on Jesus.

Warriors don't just happen. You are the result of a detailed and committed development process. You can only be as ready as your training regimen allows you to be. No night season will last forever. Every storm passes. However, the time to prepare yourself for those battles is now.

In 2 Samuel 22, King David reveals how you too can own the night when it comes your way. He says,

> "With Your help I can advance against a troop; with my God I can scale a wall… It is God who arms me with strength and keeps my way secure… He trains my hands for battle; my arms can bend a bow of bronze."[xxv]

In this Christian journey, we never outgrow our need for Jesus! If Jesus was no longer part of your life, how long would it take you to notice? If the Holy Spirit stopped leading and guiding you, would it have any effect on the life you lead on a daily basis?

If you are in the midst of a night season, invite Jesus to lead you out of it. But don't stop there! The next step to encountering a breakthrough is to crack open your Bible and read Psalm 91 today. Spend some time in prayer. Tell another strong Christian in your world about where you feel stuck, and ask them to help keep you moving forward.

Don't give up. Keep those attributes as a regular part of your daily habits. Change is possible!

Your night season will improve, but it won't happen immediately. You must stand watch in the night if you desire to see a new day. The eleventh general order of the United States military says:

> "To be especially watchful at night, and during the time for challenging, to challenge all persons on or near my post, and to allow no one to pass without proper authority."[xxvi]

Do not let your guard down when you are in a night season. Be like the sentry who has an increased level of attentiveness when on watch in the middle of the night. Allow that experience from your military service to teach you how to guard your heart and mind in the tough moments of life.

Challenge those thoughts, fears, doubts, insecurities, or cravings that are trying to prey on you in the midst of your valley or dark moments.

Tell them they will not have access to your

marriage, your reputation, your thinking, or any other area of your world. Own the battle space in which you find yourself, even when it seems like the odds are not in your favor!

For those who are not in the midst of a night season, remember you will only be able to fight the way you trained. Don't wait for the tough days to pursue Jesus. Take the same steps I just outlined above and build them into your life today. This will help insure you are prepared when the day of struggle comes knocking.

Remember this: tough days are inevitable. Murphy has a way of knocking at everybody's door sooner or later. Warriors don't cave. They don't fold. They call on Jesus. They step up and do the work that needs to be done. No matter what you face, our God is greater. We can rise above because, with Jesus, *we own the night*!

Chapter Four: ALWAYS FAITHFUL

Semper Fidelis. These Latin words mean "Always Faithful," and this motto has helped shape the fighting spirit of one of the world's most storied and celebrated military traditions: the United States Marine Corps. One of the defining characteristics of a true warrior is his or her faithfulness. The odds do not sway their commitment or willingness to fight.

We are not called to be faithful only when it is convenient, popular, understood, or beneficial. No matter the state of affairs around us, we cannot back down from our faithful allegiance to Christ!

From a natural point of view, following Jesus does not always seem like the most practical or beneficial route. That's ok because we are called to live by faith! That is why the Bible repeatedly reminds us that faithfulness is a choice.

Sadly, the world does not have much to say about faithfulness. Our society has more to say about leaving for better "opportunity" than it does about the lasting influence that can only be found through choosing to stay faithful. When I think of faithfulness, a catalog of great verses comes to mind:

2 Chronicles 16:9- "For the eyes of the Lord range throughout the earth to strengthen those whose hearts are fully committed to Him..."

Proverbs 28:1- "The wicked flee though no one pursues, but the righteous are as bold as a lion."

1 Corinthians 15:58- "Therefore, my dear brothers and sisters, stand firm. Let nothing move you. Always give yourselves fully to the work of the Lord, because you know that your labor in the Lord is not in vain!"

2 Timothy 1:7- "For the Spirit God gave us does not make us timid, but gives us power, love and self-discipline."

Hebrews 10:35, 36, 39- "Therefore, do not throw

away your confidence, which has a great reward. For you have need of endurance, so that when you have done the will of God, you may receive what was promised… but we are not of those who shrink back to destruction, but of those who have faith to the preserving of the soul."

Revelation 2:10- "…Be faithful, even to the point of death, and I will give you life as your victor's crown."

Those are the verses that immediately came to mind. However, I assure you the list could continue. If you feel your commitment to a faithful life wavering, I encourage you to put those verses wherever they can be seen on a daily basis. *Never Forget: Faithfulness is not a feeling, it's a choice!*

I wrote earlier about the general orders of the U.S. military. General Order Five speaks to the topic of faithfulness. It says a service member can only leave his or her post when properly relieved.[xxvii]

Faithfulness is about not leaving the battle too

soon. Faithfulness is about staying true even when it hurts. Faithfulness is about prioritizing loyalty above pragmatism.

I believe this trait resonates with most, if not every, individual reading this book. Your time in military service has likely revealed that faithfulness requires more than good intentions. The problem is faithfulness, like every chain, is only as strong as its weakest link.

I don't know about you, but I've got some weak links in my life! The good news is the Bible is filled with people who felt the exact same way. You don't have to fake perfection in order to receive God's attention. So what should we do?

The answer is simple: run to Jesus! Our ability to walk in faithfulness is rooted in our ability to rest in His faithfulness. With Jesus as our hope, we have no reason to be unfaithful. In Him, we find a security and a certainty that the world cannot give or take from us.

The author of Hebrews is speaking of Jesus when

he writes, "we have this hope as an anchor for the soul, firm and secure…"[xxviii] I love what this teaches us about Jesus: He is our anchor. He is firm. He is secure. Whenever we run to Jesus, we run to a faithful savior who can give us the strength to weather any storm!

In Hampton Roads, we are proud of the fact that our region is home to the largest navy base in the world.[xxix] Over the years, I've had the opportunity to be aboard a number of ships in our nation's naval fleet. These machines and their crews are awe inspiring.

Have you ever seen the anchors on those ships? The word "massive" does not adequately describe them. The largest anchors of the fleet can be seen on the aircraft carriers.

Each anchor weighs thirty tons. Each link of the chain holding the anchor weighs 360 pounds. Fifty-seven of those links forms what is called a shot of chain.

This means that one shot weighs approximately

20,500 pounds. If you want to safely secure the anchor on a carrier, it requires twelve shots of chain. This equals approximately 246,000 pounds of iron.[xxx]

When you understand the strength of these chains, it gives you an understanding of just how large aircraft carrier anchors actually are.

Why take the time to make such a large anchor? Because when the commanding officer gives the order, he must have confidence the ship will be able to remain in place. This must be possible regardless of the state of the seas or other factors surrounding the carrier. A great ship requires a great anchor.

It is God's will that we set out into the deep waters of life using the most secure anchor of all, Jesus. While He has called us to great things, He never created us to do them without Christ at work within us. If you want to sail the open oceans of life, make sure you aren't doing it alone!

You are called to live a life that stands out and shines bright for God. Big dreams lead us into big seas, and big seas usually bring big storms. Wherever life in the service leads, you have a faithful defender, anchor, and friend in Jesus. Your storm will not be allowed to have the final say!

Never Forget: You can trust Jesus.

The question is, who or what have you placed your trust in? Are you pursuing the God dream in your own strength? I've seen people start with God at the center of their pursuits. Then, after a few wins, they begin to think they are the secret to their success. All of the sudden they ditch God and press forward alone. That is the quickest way to lead your life into a ditch.

Don't make that mistake. Refuse to let your appraisal of self-worth or success to get between you and God. Pride has derailed more dreams than I can count. Don't be the next victim. We all need the strength of the Holy Spirit and the grace of Jesus leading us in our life journey.

The Bible puts it like this, "therefore let us draw near with confidence to the throne of grace, so that we may find mercy and find grace to help in time of need."[xxxi]

I love the sense of certainty in that verse. Feel free to check the original Greek if you like… there is no fine print to that promise! My prayer for you is that you would know those words are true for your life today. We are all invited to approach God with confidence.

Why come to Him? Because when we come to God, we will find grace and mercy… not judgment and anger. Also, remember these gifts are for our time of need (notice the author did not say they await our time of perfection). There is a reason we call the message of Jesus Good News… it really is GOOD NEWS!

There is one other truth we must mention as we talk about faithfulness. Followers of Christ are meant to become more like Him. When people get around us, it should be a window into what the heart of God is truly like.

That is what Paul is talking about in 2 Corinthians 5 when he writes, "…we are Christ's ambassadors, as though God were making His appeal through us…"[xxxii]

Because Jesus is always faithful, we are charged with the mission of exampling faithfulness to the world around us.

Faithfulness comes down to being a man or a woman of your word. "Let your 'yes' be yes or your 'no' be no."[xxxiii] Our God does not waiver in His commitments and neither should we.

In the closing words of his first letter to the Corinthians, Paul charges the men of the church to "be alert, stand firm in the faith, act like men, be strong."[xxxiv]

Paul has given us a great roadmap to pursing faithfulness in this passage. First, we must "be alert." God is earnest in His work to equip us for faithfulness because He understands the enemy pursuing us. Do not be naïve to the fact that there is an enemy who desires to weaken your kingdom effectiveness and influence by attacking your faithfulness.

Even in the unseen places, your faithfulness matters! Are you spiritually awake and alert? That is why regular involvement in church, prayer, and the reading of the Word of God is so essential. It keeps us spiritually aware and on watch!

The second key to faithfulness is to "stand firm in the faith." As they say in the Naval Special Warfare community, "earn your Trident every day." A SEAL is challenged to live up to the standards of his community on a daily basis. Hitting the mark yesterday doesn't mean a thing today. He must show himself to be faithful to the brotherhood in the here and now. We would be wise to learn from their example. I pray God raises up individuals in the Kingdom who will catch this revelation!

For those who have been faithful in the past through difficult circumstances, I applaud you. However, know that your faithfulness in the past does not absolve you of your need to continue standing strong today.

Next Paul says to "act like men." Now, I know women are reading this book as well. Don't tune out ladies; this point is applicable to you too! Paul is saying don't act like a child. Children whine. They throw tantrums. They wait for the hand out. They need others to change their diaper, feed them, and so much more.

The call to faithfulness is the call to maturity. Paul powerfully wrote in 1 Corinthians 13:11, "when I became a man, I put childish things behind me."[xxxv] Where in your life has God challenged you to a higher level of maturity and faithfulness? Step into it, today.

Finally, Paul charged the men in Corinth to "be strong." I believe he is challenging them to walk the road of faithfulness in the same way I am challenging you now. How do we "be strong"? Trust Jesus.

When you feel like you don't have what it takes,

ask Jesus to give you the strength. How do you ask Him? The same way you would ask a friend! Be real and be direct. I assure you He can handle it.

On deployment, stay faithful to your spouse. Whether somebody else ever knows what you do does not matter. You gave your spouse your word at the altar the day you got married. Stand by your word!

In your service to country and fellow service members, be ever faithful. Bring the best you to your command. People are counting on you to be excellent!

As a parent, faithfully love and raise your children. In the same way God leads us with patient strength, let's lead our children.

I challenge you to take a moment and ask the Holy Spirit to reveal a few areas in life where greater faithfulness is being asked of you. When He reveals it, ask Him to strengthen you and equip you to walk it out.

God never calls us to do a work without also giving us the strength, resource, resilience, and courage needed to fulfill the task!

Finally, tell some fellow Christians in your world where you are pursuing greater faithfulness. Ask them to pray for you in the days ahead and to be a supportive voice in your life. Invite them to challenge you to not back down when the going gets tough. With Jesus, you are up for it!

CHAPTER FIVE: TO LIBERATE THE OPPRESSED

Purchasing a used car is a unique experience. It requires great attention to detail from the buyer. If you aren't careful, you can find yourself with a car that has all the dents and scratches repaired but with an old, busted transmission hiding inside.

I recently heard about a home buyer who purchased a new home. Unbeknownst to him, some of the copper wiring in the house was stolen during construction. Here's the crazy thing: the builder knew that it happened. In an attempt to cut costs, he decided to just drywall over the damage anyway so that it looked like nothing ever happened.

The homeowner moved in and it wasn't long until the issues became evident and the builder's negligence was found out.

It breaks my heart to see it, but so many people treat Christ and His ability to change our lives the exact same way. Scores of people use the cross to simply renovate or decorate their lives instead of allowing it to change them from the inside out. Before we pass judgement on the builder mentioned earlier, let's be sure to look in the mirror. Are we guilty of doing the same thing?

God has always had so much more for His people. He does not pretend that life did not steal from you, or that your past doesn't hurt. He desires to truly restore, renew and free us of those painful chapters from the past. The God we serve brings genuine liberation and the message we carry is one of transformation!

Don't believe me? Take a moment and read the words that our God spoke as He set out to liberate the People of Israel from their slavery in Egypt:

"Say, therefore, to the sons of Israel, 'I am

> the LORD, and I will bring you out from under the burdens of the Egyptians, and I will deliver you from their bondage. I will also redeem you with an outstretched arm and with great judgments. Then I will take you for My people, and I will be your God; and you shall know that I am the LORD your God, who brought you out from under the burdens of the Egyptians. I will bring you to the land which I swore to give to Abraham, Isaac, and Jacob, and I will give it to you *for* a possession; I am the LORD.'"[xxxvi]

Can you sense the absolute transformation and liberation God intends for His people through His promises? God is still in the business of liberating people.

After His amazing work in Egypt, one of the most common ways He would later refer to Himself is "...The Lord your God who brought you up out of Egypt..."[xxxvii]

The Church represents Christ on this earth. As

previously mentioned, Heaven's call of salvation is sent into the world through the Body of Christ![xxxviii]

This means we are not only called to receive liberation. Having experienced the freedom that is only found in Christ, everything about our lives now boils down to bringing that same liberation into the lives of those in the world around us.

The 1st Special Forces Command of the United States Army is commonly referred to as "The Green Berets." The men serving in this command are highly specialized in their training and fall under the leadership of the United States Special Operations Command. This unit is officially tasked with five primary missions that our nation will call upon them to undertake in times of war and uncertainty: unconventional warfare, foreign internal defense, special reconnaissance, direct action, and counter terrorism.[xxxix]

The motto that shapes their ethos and esprit de corps is the Latin phrase "de opresso liber" which means *to liberate the oppressed.* This brave community and the men and women who support them are the consummate "silent warriors."

They understand the freedom loving people of America have not invested in and trained them for petty bragging rights or a good bar room story. They have been entrusted with the lethal secrets of waging war so that they may liberate those who can't liberate themselves.

I believe this motto should describe the mission of the church as well. Jesus explained His reason for coming to this earth when He plainly stated, "For the Son of Man came to seek and to save the lost."[xl] In my years of pastoring in Virginia Beach, I have had the chance to sit and bring clarity in the midst of challenging times to many serving our nation.

I have often been told that deployment can feel monotonous: Do one's duty. Workout. Eat. Sleep. Repeat. Over and over again. One day is the same as the one hundred before it. What if I told you that God has so much more for you to accomplish in this season of your life than that? What if there is more to you than your rate?

What if a deployment was not only a season of being overseas and defending America? What if it was also a season of being sent by God into our world as a missionary? Wherever you go, you can bring Jesus with you. What I love about Jesus is that our natural limitations cannot limit His ability to use us. He is willing and able to send you into this world for Kingdom influence.

Perhaps you feel limited by your pay, your education, your family origin, your rank, or any other number of things. There is an incredible account about a healing in the ministry of Peter and John during the early days of the church. I believe it applies to this idea we are now exploring:

> "When he [the cripple] saw Peter and John about to go into the temple, he *began* asking to receive alms. But Peter, along with John, fixed his gaze on him and said, 'Look at us!' And he *began* to give them

> his attention, expecting to receive something from them. But Peter said, 'I do not possess silver and gold, but what I do have I give to you: In the name of Jesus Christ the Nazarene—walk!' And seizing him by the right hand, he raised him up; and immediately his feet and his ankles were strengthened."

Peter and John knew what they did not have access to in their season, but they had an even greater knowledge of what they did have access to. And their knowledge of what they had in Jesus enabled them to bring liberation to that person on the side of the road that day.

Perhaps our greatest obstacle to bringing liberation to the world around us is that we have lost our sense of urgency. How urgent is your desire to see the people in your unit, your family, and your world come to know Jesus?

Have we forgotten what it is like to live without Jesus and His promises at the center of our lives?

› never forget there are still people living
..ng marriages, raising kids, and trying
.. discover their life purpose without the power and presence of the Holy Spirit. If we truly believe what the Bible teaches about Heaven, Hell, and eternity than it should instill an overwhelming urgency in each of us!

Jesus said, "We must work the works of Him who sent Me as long as it is day; night is coming when no one can work."[xli] Jesus lived with an urgent focus. Do you?

The Holy Spirit has given us many great gifts (to learn more, read *Overdrive: Life, Led by the Holy Spirit*). They weren't given so we can simply feel good about ourselves. They have been entrusted to us so we can help liberate others in the same way that we have been liberated.

A simple test of our urgency and focus is to audit our prayer lives. How much of your prayers are about you? How much are they about others? Are we praying for parking spots or nations? Let's be people who pray often for the world around us.

In the old days, people formed lines from a water source to places that had caught on fire. Then, they would pass buckets from one person to the next so the water could quickly travel from the well or river to the scene of the fire. Now, let me ask you this: have you been saved? Have you been set free? Has God worked a miracle in your marriage or family? If so, praise God!

But know this, the time has come for you to bring that "bucket" we were just talking about to the fires in the lives of others in the world around you. Do not turn your back on the urgency you encounter—ask Jesus to give you a heart for the people in your world who don't yet know Him!

What good is freedom if we have nobody else to share it with? I assure you, seeing my children walk in relationship with Jesus is far more fulfilling than when Christ first found me all those years ago. Church is amazing, but it is an entirely different adventure when you include your family and friends!

There is more in store for you, but you'll never

find it looking after yourself. When we commit to liberating and leading the people in our world to Jesus, everything changes. Only then will we encounter the adventure and blessing hidden in the Christian life.

Long before Jesus walked this earth, one of Israel's prophets penned words to help his people. He wrote them to reveal the motives and works of the promised Son of God. The words below are from Isaiah 61. To this very day, they reveal what Jesus desires to do in the hearts of people just like you and me.

> "The Spirit of the Lord GOD is upon me, because the LORD has anointed me to bring good news to the afflicted; He has sent me to bind up the brokenhearted, to proclaim liberty to captives and freedom to prisoners; to proclaim the favorable year of the LORD and the day of vengeance of our God; to comfort all who mourn, to grant those who mourn *in* Zion, giving them a garland instead of

> ashes, the oil of gladness instead of mourning, the mantle of praise instead of a spirit of fainting. So they will be called oaks of righteousness, the planting of the LORD, that He may be glorified."

We can have that same influence on the people in our world. We can liberate the oppressed, those who are living this life without the love and promises of God. Often, I believe we make liberating the oppressed much harder than it needs to be.

God can use you right where you are, just the way that you are. How? Just ask. Ask Jesus to use you. Ask Him to show you the hurting people around you. Ask Him to let you hear the cry for rescue. But here is the catch: my God answers prayer! If you pray these prayers, you need to be ready.

It won't be long until that bucket of water that we talked about earlier will be passed to you, and it will be your time to bring the "water" of the Gospel to the people in your world.

The Gospel cannot be contained! It doesn't matter if you are on the home front or the battlefront. Whether in a sub, on a carrier, in a FOB, or in an airport traveling to far-away places Jesus can and will use you.

Before you go any further, stop and ask Him to use you to free the people in your world! Write down the names of three individuals in your life and ask Jesus to reveal to you how you can use your interests, skills, knowledge, or talents to reach them with His love. I promise, your life will never be the same!

Chapter Six: SEND ME

Over the course of the past five chapters, we have discussed quite a few things. However, we both know the world is not changed through discussion. It is changed through action.

It is with that thought in mind we now return to the very same passage of scripture where this book began, Isaiah 6:8 (NIV), "Then I heard the voice of the Lord saying, 'Whom Shall I Send? And Who will go for us?' And I said, 'Here am I. Send me.'"

My prayer is that this book would impact lives in the most powerful way imaginable. I pray this text disturbs the comfortable and comforts the disturbed.

I hope it proves to be impossible for you to return to your national service and continue doing things the same way you did them before reading this book. I am hopeful that the revelations this book brings will forever change how you see yourself and your fellow service members. You truly are a community of influencers set apart for a Kingdom purpose. Never doubt that!

Who you are and what you do is not only a matter of national significance. You are charged by our Creator to have an eternal impact on the people in your world! The question now is will you go? While the need is very real, our God is a gentleman. He will not force you into being a participant in His plans and purposes. You have a choice to make. He will not send you into Heaven's fight without your agreement.

This reality brings to mind the fighting spirit and heritage of our United States Army Ranger Regiments. The Ranger Creed underscores the values which set these fine American warriors apart. The opening lines boldly declare,

> "Recognizing that I volunteered as a ranger, fully knowing the hazards of my chosen profession, I will always endeavor to uphold the prestige, honor, and high

esprit de corps of my ranger regiment."[xlii]

Allow me to draw out a Kingdom principle from this creed: have you volunteered yourself to be used by God for His Kingdom and purposes?

Some people think being used by Jesus is like waiting for your number to be called at the DMV. Until your number is called, all you can do is sit in an uncomfortable seat and watch the hours of your day tick by. Nothing could be further from the truth! *We choose to be sent.* Right here and now God has a mission for you… if you will take it.

The first key to being used by God is refusing to ignore the brokenness we see on a daily basis. Proverbs passionately reveals what happens when we pretend to not see the needs unfolding in the world around us:

> "Rescue those being led away to death; hold back those staggering toward slaughter. If you say, 'But we knew nothing about this,' does not he who weighs the heart perceive it? Does not he who guards your life know it?

Will he not repay everyone according to what they have done?"[xliii]

The first key to being sent: If you see something, do something!

Pretending to not see what's wrong with the world around us will not cause humanity's problems to conveniently disappear until a more ideal time. We are now responsible. Just as you are held accountable by your commanding officer for fulfilling your duties, so too Christians will be held accountable for what we do with this gift of life.

Eternity is real! Yet we live in a world that conditions us to focus all of our efforts on this present life. Society tries to blind us to what matters most with the distractions of the here and now. Don't fall for that trap. As long as this life seems, I assure you the time each of us spends in either Heaven or Hell will be far longer.

The second key to being sent: Refuse to live distracted!

Jesus put it like this when he was talking about pursuing God and His purposes for our time on this earth:

> "Don't look for shortcuts to God. The market is flooded with surefire, easygoing formulas for a successful life that can be practiced in your spare time. Don't fall for that stuff, even though crowds of people do. The way to life — to God! — is vigorous and requires total attention."[xliv]

Jesus said "crowds" of people have been distracted and led off course from the path that leads to true life. Are you a part of the crowd He was talking about? Have you gotten used to living distracted?

Common distractions that rob people of the great purposes Jesus has for them are past regrets, fears, doubts, insecurities, and comparison just to name a few.

One of my favorite hobbies is riding motorcycles. There is nothing else out there quite like it. As enjoyable as it is, it doesn't come without risk. Riding distracted is never a good idea. One of the quickest ways to crash is to focus on where I don't want to ride instead of looking where I do want to go!

When on a winding backcountry road, I have learned to look through the turn to where I am headed instead of constantly glancing at the ditches on either side of the road. Life with Jesus is the same way!

I encourage you to look through the turns of life. If we continue to focus on the distractions, we will only end up more confused. Focus on Jesus and who He says you are. Focus on what He said you can do!

Jeremiah 33:3 is a fantastic verse for those distracted seasons of life: "Call to me and I will answer you and tell you great and unsearchable things you do not know."[xlv]

Feeling distracted? Call out to Jesus. He will recalibrate your focus and motives.

The third key to being sent: Open "comms" with God!

If you are going to discover God's best for you, than you must have an open line of communication with God. Think about the value of clear communication in combat. It can be the difference between life and death!

People seriously overcomplicate communication with God. In Jeremiah 33, He tells us that if we will simply call out to Him, He will speak and reveal His will to us.

It's easy to find ourselves calling out to the wrong things in life. Many call out to the media, popular opinion, the stock market, emotions, or personal performance to find their life's identity and purpose. All of those things can change or be destroyed in a moment.

The Bible reveals that God is unchanging. In a world that is always changing, we can have a rock solid sense of purpose and value, but we only find it when we call out to God!

The fourth and MOST IMPORANT key to being sent: Know Jesus!

Everything we have talked about in this text rises and falls on a personal relationship with Jesus. Please note that I did not say "knowing about Jesus." I said knowing Him. There is a world of difference between the two!

I know about our military. I stay up to date about what the news says concerning our brave men and women in the armed forces. I count it an honor to have close friendships with many veterans and active duty members of our military.

I've even read many books about the most celebrated operations in American military history, but here's the thing, I've never worn the uniform. Knowing about the operation and being a part of the operation are two completely different things. Knowing the facts, terminology, and tradition does not mean I am a part of the community. It just means I know about it.

Knowing what the Bible says about Jesus, what

your family told you about Jesus, or what pastor preaches about Jesus are all well and good… BUT it does not mean you know Jesus for yourself!

I challenge you to open your life to Jesus today. Wherever you are and whatever you are doing, if you have never asked Jesus to be a part of your life than you can do it right now! Simply pray these words:

"Jesus, I desire to know You. I want the greater life that You have for me, but I cannot live it on my own. Help me. Forgive me of my past sins, and teach me to be like You. Thank You for loving me. I receive Your gift of forgiveness and choose to follow You. I am now a Christian. Amen!"

Paul boldly wrote that if anyone is in Christ, they are a brand new creation![xlvi]

You may look the same on the outside, but God has begun the exciting process of changing you from the inside out. If you just prayed that prayer, congratulations! Your life will never be the same!

Your next step is to become connected to and

involved in a healthy local church.

If you need help finding a local church or would like to celebrate the decision you just made, I encourage you to write us at military@wavechurch.com. It would be a great honor for us to come alongside you and help you figure out your next steps in your journey with Jesus.

If you've known Jesus for a while, perhaps your first step to being "sent" in your current season would be to share this book with somebody in your unit.

For some of you, being sent will mean re-engaging with your wife and children and pressing through the difficulties and pains of recent chapters in your family life. For others, it could be taking off the "mask" of perfection and letting a fellow Christian know about a struggle in your life and asking them to help hold you accountable as you pursue the greater life that

Christ has for you.

Know this: there is a voice still crying, still asking "who will go for us?" Today, if you are willing, God can and will send you. His passion is to send you on a mission; a great adventure. A mission unlike any other. God's mission is for you to liberate the captive, bring justice for the oppressed, and establish a legacy for future generations that is unmistakable and unshakeable.

Introduction

[i] US Army. (n.d.). Oath of Enlistment. Retrieved September 1, 2017, from https://www.army.mil/values/oath.html

[ii] Revelation 3:15-16 (NIV)

[iii] Lewis, C. S. (2012). What lies behind the law. In *Mere Christianity* (p. 45). New York: HarperOne.

[iv] Romans 13:1 (NIV)

[v] Acts 10:2 (NIV)

[vi] Exodus 15:3 (NIV)

Chapter One

[vii] Hebrews 12:2 (NIV)

[viii] Groman, D., & Christensen, L. W. (2013). *On Combat: The Psychology and Physiology of Deadly Conflict in War and in Peace*. New York: Hachette Audio.

[ix] Spielberg, S. (Director). (1998). *Saving Private Ryan* [Motion picture]. USA: Dreamworks Pictures.

[x] Isaiah 40:29-31 (NIV)

[xi] Proverbs 28:7 (NIV)

[xii] Habakkuk 3:19 (NIV)

[xiii] US Army. (2017). Guide to the Wear and Appearance of Army Uniforms and Insignia. In *Guide to the Wear and Appearance of Army Uniforms and Insignia* (p. 234).

Chapter Two

[xi] Chesterton, G. K. (1911, January 14). *The Illustrated London News, CXXXVIII*(3745).

[xv] US Air Force. (n.d.). Enlisted Pararescue. Retrieved September 1, 2017, from https://www.airforce.com/careers/detail/pararescue

[xvi] John 15:13 (NIV)

[xvii] 2 Timothy 4:2, (NIV)

[xviii] 2 Timothy 2:3,4 (NIV)

[xix] John 3:30, (NIV)

Chapter Three

[xx] Suarez, G. (1999). *The Tactical Rifle: The Precision Tool for Urban Police Operations*. Boulder, CO: Paladin Press.

[xxi] Luke 10:2 (NIV)

[xxii] US Army. (n.d.). 160th SOAR (A). Retrieved September 1, 2017, from http://www.soc.mil/USASOAC/160th.html

[xxiii] Psalm 91:1-2, 4-6 (NIV)

[xxiv] Von Clausewitz, C. (1989). *On War: On victory and defeat* (Princeton shorts ed.) (M. Howard & P. Paret, Comps.)

[xxv] 2 Samuel 22:30, 33, 35 (NIV)

[xxvi] US Navy. (n.d.). General orders. Retrieved September 1, 2017, from hhttp://www.bootcamp.navy.mil/general_orders.html

Chapter Four

[xxvii] US Navy. (n.d.). General orders. Retrieved September 1, 2017, from http://www.bootcamp.navy.mil/general_orders.html

[xxix] Hebrews 6:19 (NIV)

[xxix] US Navy. (n.d.). Naval Station Norfolk. Retrieved September 1, 2017, from https://cnic.navy.mil/regions/cnrma/installations/ns_norfolk.html

---

[xxx] US Navy. (2015, May 14). Pulling their weight: anchor and chain. Retrieved September 1, 2017, from http://www.navy.mil/ah_online/ftrStory.asp?id=87091

[xxxi] Hebrews 4:16 (NASB)

[xxxii] 2 Corinthians 5:20 (NASB)

[xxxiii] Mark 5:37 (NASB)

[xxxiv] 1 Corinthians 16:13 (NASB)

[xxxv] 1 Corinthians 13:11 (NASB)

Chapter Five

[xxxvii] Exodus 6:6-8 (NASB)

[xxxvii] Psalm 81:10 (NASB)

[xxxviii] 2 Corinthians 5:20 (NASB)

[xxxix] US Army. (n.d.). United States Army Special Forces Operations Command. Retrieved September 1, 2017, from http://www.soc.mil/usasfc/hq.html

[xl] Luke 19:10 (NASB)

[xli] John 9:4 (NASB)

Chapter Six

[xlii] https://www.goarmy.com/ranger/about-the-rangers/ranger-creed.html

[xliii] Proverbs 24:11-12 (NIV)

[xliv] Matthew 7:13-14 (MSG)

[xlv] Jeremiah 33:3 (NIV)

[xlvi] 2 Corinthians 5:17 (NIV)

ABOUT THE AUTHORS

Steve Kelly is the Senior Pastor of Wave Church, a Christian church with national and international influence. Wave Church has campuses in Virginia, North Carolina, and California-- with more to come.

Pastor Steve's passion is to win the lost and to build the local church. Wave Church's primary mission is to help people discover their purpose in Christ and to make a difference through being planted in the local church.

Joe Riddle is the campus pastor of Wave Church's Seaboard Road location and oversees Wave's military ministry initiatives. Joe's best decision in life was asking his bride, Lauren, to marry him. His proudest accomplishment is raising his two sons and daughter. Joe refuses to miss any opportunity to pick heavy stuff up and to put it down.

23981581R00050

Made in the USA
Columbia, SC
16 August 2018